Phantom Bouquets

Phantom Bouquets

Historical and Modern Methods for Skeletonizing Leaves

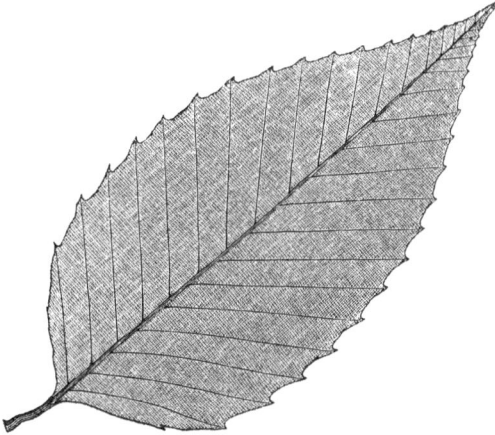

Chad Arment, Editor

COACHWHIP PUBLICATIONS
Landisville, Pennsylvania

Phantom Bouquets: Historical and Modern Methods for Skeletonizing Leaves
Editor, Chad Arment
Copyright © 2008 Coachwhip Publications

ISBN 1-930585-64-0
ISBN-13 978-1-930585-64-5

The Phantom Bouquet: A Popular Treatise on the Art of Skeleton-izing Leaves and Seed-Vessels and Adapting Them to Embellish the Home of Taste, by Edward Parrish. First published in 1862 (Philadelphia, Pennsylvania: J. B. Lippincott & Co.).

"Phantom and Skeleton Leaves," from *Treasures of Use and Beauty: An Epitome of the Choicest Gems of Wisdom, History, Reference and Recreation*. Multiple authors, this section anonymous. Published in 1883 (Detroit, MI: F. B. Dickerson & Co.).

Cover image: Sapsiwai (Fotolia)

Coachwhipbooks.com

Contents

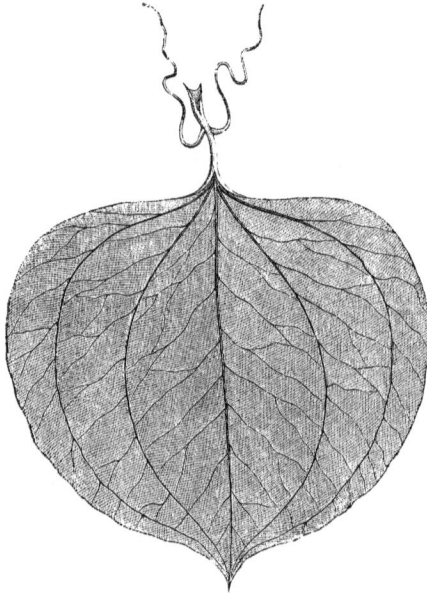

Fig. 1. Green Briar Leaf,
the under surface, showing the venation
of an endogenous leaf.

The Phantom Bouquet
(1862)

Edward Parrish

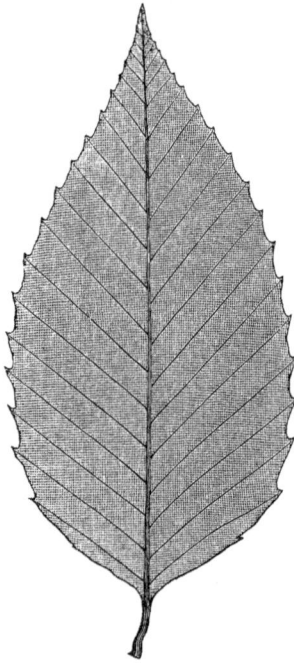

Fig. 2. Leaf of the Beech Tree,
showing midrib and lateral veins,
present in most exogenous leaves.

Preface

This essay was written in the autumn of 1861 for the *Atlantic Monthly*, and accepted for publication by the editors of that popular Magazine; but the pressure upon its pages has prevented the appearance of an article which is so little in accordance with the tone of the current American literature during the past eventful year. The numerous applications to the author for instructions in the art of Skeletonizing have induced the determination to delay its publication no longer, and to change it from a magazine article to a small practical work, adapted to aid the tyro in attaining a perfect acquaintance with the subject of which it treats. It is hardly necessary to acknowledge the aid derived from numerous friends of both sexes, who have freely imparted the results of their experience in the matter in hand. Of course, all skeletonizers have learned by this time that it is only by communicating what they know that they can expect to receive in turn the ideas of others, and thus promote our beautiful pursuit to its true position among the liberal arts.

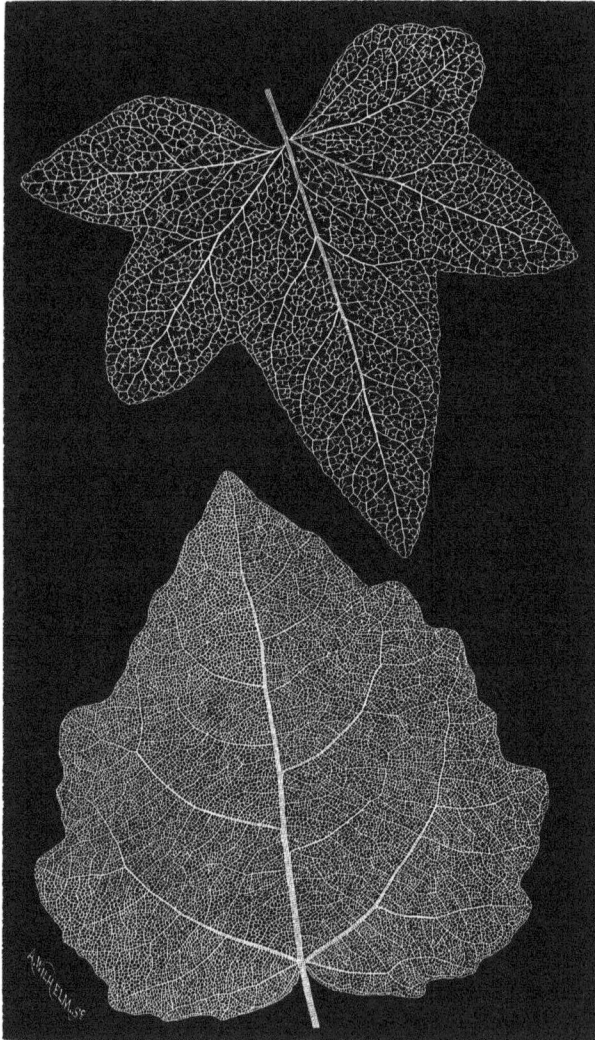

Fig. 3. Ivy; Silver Poplar.

Historical

Some years ago the writer was attracted by a beautiful vase of prepared leaves and seed-vessels, displaying the delicate veinings of these plant-structures deprived of their grosser particles, and of such brilliant whiteness as to suggest the idea of perfectly bleached artificial lace-work or exquisite carvings in ivory. This elegant parlor ornament was brought by returning travellers as a novel and choice trophy of their transatlantic wanderings: none could be procured in America, and no one to whom the perplexed admirer could appeal was able to give a clue to the process by which such surprising beauty and perfection of detail could be evolved from structures which generally rank among the least admired expansions of the tissue of the plant. That the novelty of this spectacle then constituted one of its attractions need not be denied; for who that has learned to dwell familiarly on any object of unusual beauty, but can still recall the emotions of delight it created when for the first time it attracted the unaccustomed eye? Yet the "Phantom Case," now that hundreds of pier-tables and étagères in city and country are garnished with its airy forms, and its photographic miniature, under the well-chosen motto, "Beautiful in death," is displayed in almost every stereoscope, still delights with a perennial charm, creating a desire, among all amateurs in matters of taste, to add an ornament so chaste to their household treasures. To this end, an unpretending though sincere lover of nature proposes to lay before his fellows of that genial fraternity which knows neither sex nor nation, a simple and easy art, which, while it will prove a pleasurable addition to the arcana of home-occupations, will in its results add to the tasteful embellishments of the household.

11

Reader, suppose not that this elegant art for which we have
no more elegant name than Skeletonizing, is any thing new
under the sun. Place it rather among the lost arts revived; for
among the quaint old curiosities to be found in the houses of
retired sea-captains and East India traders you will often find
Chinese pictures, sometimes of considerable beauty and ingenuity, exhibiting flowers, fruit, shells, birds, or insects painted
in bright colors on veritable skeleton leaves. Moreover, some
of the old London books we have lately forgotten to read give
accounts of the identical process, and tell us that, as long ago
as 1645, Marcus Aurelius Severinus, professor of anatomy and
surgery at Naples, turned his attention to the subject, and published a figure of a leaf thus delicately prepared. But this inge-
nious disciple of Æsculapius, according to the fashion of his
time kept the process a secret, and so we owe probably the first
published account of the method of preparing plant-skeletons
to a Dutch naturalist, Fredrick Ruysch by name, who in 1723
first gave to the world the announcement that, through the
putrefactive fermentation promoted by warmth and moisture,
the pulpy matter of the leaf may be loosened so as to be separated from the fibrous skeleton, which may thus be preserved
unimpaired. This, reader, is the secret which has for the past year
attracted so much attention in the neighborhood of Philadelphia: it was at first communicated in under-tones by certain
confidential lady amateurs to their amateur friends, with scru-
pulous injunctions of secrecy; but now many observations and
experiences have been exchanged, and in this charmed circle
the new art has attained to greater perfection than could have
been expected in a single season.

An incidental but important result of the cultivation of this
elegant pursuit is an increased attention to leaves. Many to
whom a leaf was formerly an object of not the least interest,
except perhaps in its connection with the beauty and effect of
the shade-tree, or as adding to the variety of the landscape, now
eagerly examine it in detail, with reference to its adaptation to the
phantom case, and study its outline, whether serrated or entire
on its margin, divided, lanceolate, ovate, acuminate, cordate,

or irregular in shape. Ferns and lycopods are carefully searched for, with a similar end in view; and, strange to say, the somewhat unsightly burs, the persistent calyxes, and dry and indurated seed-vessels of our wayside weeds, are invested with a new interest by their suspected or ascertained fitness for projected skeleton cases. Since the past summer, hundreds of these have been made tributary to the most chaste and refined ornamentation.

Something of Science

Without inflicting on the reader a misplaced disquisition on Botany, or its correlative science, Vegetable Physiology, it may be here explained that the leaf, considered in its physiological relations, is to the plant what the digestive and breathing organs are to the animal. Its porous structure permits it to be permeated by moisture drawn up from the soil through the root and stem, and also by the air in which it so incessantly vibrates. These elements, charged with principles adapted to the nutriment of the plant, meet in the leaf-tissue, where they are metamorphosed into the true elements of growth, and elaborated into new cells fitted to develop and increase the living and growing plant. By what subtle influences the sun's ray, acting on these gaseous and liquid particles, works to such marvellous ends, producing the innumerable vegetable forms, from the slender grass to the massive oak, and all the infinite variety of fruits of the field and flowers of the meadow and garden, Science has in vain sought to answer. Only the microscope, through its mysterious revealings, points to the cell as the initial, almost infinitely minute, alembic in which this inscrutable and beneficent synthesis is ever going on.

What the older botanists were wont to call nerves, but the modern have named veins, are spread more thoroughly over the leaf than would appear from an examination of the surfaces: in fact, the loosely distributed cellular tissue is everywhere traversed by these ducts, conveying the sap to the extremities, there to be subjected to the subtle chemistry of the cell. The mode of distribution of these veins is one of the leading characteristics observed in the classification of plants. The two great divisions of

Endogens and *Exogens*, having distinct methods of growth and distinct structure of the seeds, differ by a distribution in the one case of the veins in curved lines from the origin to the apex of the leaf, while in the other and far more numerous and important class they diverge for the most part at regular angles from a central stem, called the midrib, to the margin. (See figures 1 and 2.) That this distinction is not artificial, or invented to detract unnecessarily from the uninterrupted gradations of plant-structure, will be apparent to any observer who will take note of the peculiarities of each of these great classes. To the skeletonizer it will be interesting to know that very few of the Endogens have firm enough structure in their veinings to furnish specimens for his art: only two or three are known to the writer, of which the common climbing green brier (*Smilax rotundifolia*) and wild yam (*Dioscorea villosa*) are the most familiar.

The Leaf a Type of the Tree

The veinings of the leaf exhibit striking evidence of that unity of design, combined with special adaptations, displayed in every department of nature.

What the trunk and branches are to the tree, are the veins to the exogenous leaf. By continuity and compactness of structure, the delicate spiral fibres which constitute the veins are endowed with strength and elasticity adapting them to sustain and supply the loose tissue which fills up the interstices between them, while, obeying the primal law of growth stamped upon the parent tree, each separate leaf spreads its continuous skeleton into a flattened outline of the tree itself.

A glance at the well-developed tree by twilight, stripped of its leaves, with nothing to obstruct its figure against the sky, will scarcely fail to recall, to one accustomed to observe and to compare, the general outline and arrangement of the skeleton leaf. Nor is this a mere fancied resemblance. Dr. McCosh, of the Queen's University, Ireland, in his comprehensive work on "Typical Forms and Special Ends in Creation," has demonstrated the correspondence between the disposition and distribution of the branches of the tree and of the leaf-veins. The very angles at which the branches leave the trunk are shown to correspond in many individual cases with those formed by the lateral veins and midrib upon the leaf; and even the curves which give grace and contour to the tree are repeated among the veins which permeate its leaves. This is believed to be a great natural law throughout the almost infinite variety of vegetable forms.

We rarely see trees which have not been somewhat changed from their normal mode of growth by prevailing winds, or too partial sunlight, or by cattle browsing on them, or man pruning or otherwise mutilating them: yet so far as the true pattern of the whole plant has been studied, it is found to correspond beautifully with that of its every single leaf.

Observe, also, how often the length of the petiole, or leaf-stalk, has a direct relation to the height reached by the stem in its unobstructed growth before branches strike off from it. In the case of the box, the privet, the laurel, the snowberry, and other favorite lawn or hedge plants, and the oak, the elm, the beech, and some other common forest trees, which when unpruned incline to send out branches from the very base, the leaf, in botanical language, is almost sessile, sitting directly upon the stem; while those which have naturally a bare trunk, such as the cherry, the apple, the pear, the chestnut, the poplar, and, above all, our splendid American tulip poplar tree, have leaf-stalks of greater length, sometimes exceeding that of the leaf itself.

Without designing to pursue this somewhat abstruse subject farther than will suffice to vindicate its interest and importance and lead the amateurs of our art to trace out these relations for themselves, it will suffice, in conclusion, to direct attention to a peculiarity in the arrangement of certain divided and com-pound leaves, which accords beautifully with the idea we have been attempting to illustrate.

Our familiar sycamore or buttonwood tree has a very char-acteristic growth. Its bare trunk, rising eight or ten feet from the ground, commonly divides itself at once into four or five branches of great length, the whole aspect of the tree in its branching and leafage being loose and open; so its leaf, branch-ing out from a long foot-stalk, is sustained by five midribs, the intervening structure being extremely delicate, so much so that no skeletonizer with whom I am acquainted has succeeded in separating and preserving it. The horse-chestnut presents a similar instance. The seven branches into which its leaf-stalk divides, constituting as many separate leaflets, follow the same type as the perfect tree which spreads from the top of its bare trunk a whorl of ascending and spreading branches.

Reader, we may now be charged with having trenched somewhat upon the domain of the imagination; and if the stern logic
of unexplained or conflicting facts is brought in contact with
these inspiring ideas, sustained as they are by many observations
of undoubted import, let us not contend, but rather confess that,
like all human strivings after the archetypes of nature, this may
contain such errors as are inseparable from imperfect observ-
ing powers and the finite reason of human beings, whose
"knowledge is patchwork," after all. We may, however, derive
from such observations much pleasure; they will impart new
and unexpected interest to the objects to which they relate, and
give to the process of skeletonizing, which unmasks the structure of the leaf, a practical and a scientific importance little
anticipated by those who have pursued it with an exclusive view
to its ornamental and aesthetic results.

How and What to Collect

Skeletonizing may be out of season when this essay for the first time falls into the hands of the reader; but as soon as the summer foliage has fully expanded in luxuriant variety, the collection of material may commence. It is important that the amateur should make the proper choice of leaves. It will be observed that those which have just opened upon the fresh-grown terminations of the limbs are relatively softer, of a lighter color and more delicate fibre, than those which expanded at the first summons of the spring, in close proximity to the solid woody branches. The little suckers which are apt to grow up from the origin of the roots often bear enormous leaves, but with very little stamina.

Every leaf for our purpose should be fully formed, having attained the firmness belonging to complete maturity, and should be without a blemish. The punctures made by insects are often fatal to success when every other condition has been ful-filled. In an appendix to this essay, prepared with the aid of Dr. J. Gibbons Hunt, and several ladies of experience in the art, a list is given of those trees the leaves of which we know to have been successfully skeletonized. Almost any or all of our decidu-ous shade-trees might probably be included; and, variety as well as beauty of form being kept in view, the pretty little leaves of the box, the rose, the pear, and the apple will be as highly prized as those of the larger magnolias, tulip poplars, maples, and lindens.

The lemon, orange, camilla, caoutchouc, and other tropical trees found in our greenhouses, besides being well fitted to the process, have the advantage of being obtainable in the winter, when our own more familiar trees are bare and leafless.

No leaves need be gathered from herbs: the organs of these short-lived tenants of the meadow and copse are formed to develop only a succulent stem and soft, flabby leaf, which will soon disappear in the processes we are about to describe, leaving no skeleton to perpetuate their outlines. The various-leaved oaks are unsuited to maceration from a different cause: they have developed underneath their glossy and tough exterior a principle called tannin, which gives to the nut-gall and oak-bark their great commercial value, and which effectually obstructs the decomposition by which we propose to get at the delicate lace-work of the leaves.

Maceration

The first step in the process of skeletonizing is to arrange suitable vessels for conducting the maceration. Almost any utensils of wood or earthenware which will hold water will answer the purpose; a single vessel will serve, but there is an advantage in having several, so as to separate the rough calyxes from the tender leaves, and of leaves, to place such as are porous and rapidly decomposable apart from the comparatively tough and unyielding. The specimens are to be covered with soft boiling water, and to be held in place by pieces of glass or other suitable weights, and in the summer placed on a shed or veranda, preferably exposed to the sun during part of the day. A closet near a flue, or some other warm situation, will be better in the winter. Fresh water must be added from time to time, to compensate for evaporation, but otherwise our rather unsightly and presently offensive mass is best out of sight and out of mind. In four or five weeks the leaves should be examined, and upon those sufficiently softened to admit of cleaning, this somewhat delicate operation may be commenced.

In the case of the ivy, holly, and some other leaves, after maceration the tough skin will peel off from the two faces in separate layers, so as to expose the skeleton covered only by loose parenchyma, which readily washes off. Maples, linden, elm, horse-chestnut, abele, and others, on the contrary, have so slight an epidermis that it soon softens and decays. They require a different treatment; the decayed leaf, being laid smoothly upon a plate or pane of glass, should be swept lightly over with a soft tooth-brush or shaving-brush till cleaned, turning the leaf over at intervals. Sometimes it will be found necessary to return the leaf to the macerating vessel, to complete the softening process. Leaves of very lacey and delicate texture may be best cleaned without a brush, by washing them gently between the thumb and finer under water.

The holly is one of the most difficult leaves to clean. It softens by maceration in from six to twelve weeks, but the skin is very difficult to separate from the hard, thickened edge, and especially from the prickles: a small pair of forceps, used with dexterity, will assist in this operation. When obtained in a perfect condition, this skeleton, like the lemon, japonica, and some others, displays a complete double structure, as though two perfect leaves had been laid one upon the other and grown together at the edges. The leaves of the numerous species of oak, and of hazel and buckthorn, resist the process of maceration so persistently as to induce a resort to other means to obtain their shadowy outlines.

Other Processes

One process, as applied to oaks, consists in flattening out the leaf, previously thoroughly browned and dried by means of a hot sadiron, and then gently tapping over its entire surface with a tolerably soft clothes-brush. If conducted skilfully and thoroughly, this pricking with the sharp bristles will be successful in separating the skin and cellular portions. But skeletons prepared in this way are seldom so delicate or so perfect as those obtained by maceration, and are also so difficult to bleach thoroughly as to be unavailable for grouping with the snow-white products of maceration.

Boiling with soap completely loosens the skin of some leaves, the best instance we have seen being that of a camelia japonica,—a beautiful leaf,—the skin of which was blistered over the whole surface, so as to peel off with perfect facility, and, by washing under water between the thumb and finger, became a perfect skeleton. No allusion has been made to the use of chemical agents, or the substitution of long boiling for the process of maceration, as applied to leaves generally, because the writer has had but poor success with either, and prefers the slow and sure method already described.

By Caddice-Worms

It yet remains to notice, in connection with oak-leaves, what cannot fail to excite the liveliest pleasure in every naturalist who delights to seek the woods and streams on chill autumn days, though all the fragrant epigeas, the delicate bloodroots, the pale spring beauties, the modest "quaker ladies," and all their lovely spring companions, have so long departed as to diffuse almost a feeling of sadness in visiting the now desolate slopes they rendered so inviting. Let our amateur note what became of the leaves that, having performed their allotted part in the growth of the forest and ceased to be permeated by the life-sustaining sap, have yielded to the blast, and now thickly strew the ground, awakening, as stirred by the wind or the foot of the pedestrian, the familiar rustle of the autumnal woods. These are all destined to pass into the earth from which they sprang, by a slow but sure decay. The oak-leaves, as would be supposed, longest resist this destiny. Even those that have fallen into yonder stream have not matted themselves into the slimy mass except by mixing with other and less hardy leaves; and here, if the explorer will search closely, he may occasionally find almost perfectly skeletonized oak-leaves. How came they so? Look: provident nature has found a way to make them, intractable as they are, to subserve a purpose in her wise economy. Thousands of curious little animals, called caddice-bugs, who envelop themselves in a tubular little cocoon of pebbles and sand, are daintily masticating the soft parts of these, leaving all the veinings as perfect as the most captious skeletonizer could desire. It is true that after the rough usage of the running stream upon its pebbled bottom and the thick matrix of twigs,

24

chestnut-burs, acorns, and the like, very few perfect specimens remain; but then, my friend, here is a hint for us. Change these adverse conditions; colonize, by the aid of an exploring kettle, a few hundred caddices, with their movable tents, to your own sheltered veranda; give them a shallow dish, with a bed of sand in the bottom, and a constant trickle of fresh water to resemble their native stream; then supply them with their favorite leaf, and they will clean it for you to perfection. This has been done successfully, and it can be done again.

Seed-Vessels

Many mature seed-vessels, and the calyxes of certain flowers, may be skeletonized by maceration or by boiling. These add greatly to the interest and beauty of the phantom case; though it is an obvious fault in most of those prepared in England that they are cumbered with too many large and comparatively opaque objects, at the expense of that airy character which is one of the chief charms of these ornaments. That unsightly weed growing with rank luxuriance on almost every heap of rubbish, especially in the suburbs of all cities, — an opprobrium to both hemispheres, each of the continents disowning its origin, though all know it but too well, — the thorn-apple, or, as we call it in America, from its "discovery" in the vicinity of the first settlement in Virginia, the Jamestown weed, matures a bur in the autumn, a skeleton of which is present in almost every phantom case I have seen. This should be gathered when it has only slightly opened at the apex, then boiled for some hours or macerated for several weeks, till sufficiently softened, and scrubbed with a tooth-brush under a running stream of water till perfectly cleaned. The seeds are to be removed by shaking, and the cellular structure, called the placenta, pricked out, though without removing the beautiful ligneous structure in which it is embedded. Water should be allowed to run freely through the bur till it is perfectly cleaned, and attention should be given to the external prickles, that every particle of the skin be removed; otherwise, after bleaching, these prominent tips will be less white than the rest. So tough is the ligneous fibre in this bur that it may be freely opened and closed without breaking it or materially weakening its union with the stem.

26

Poppy-heads—the pericarps of *Papaver somniferum*—require only about two weeks' maceration. As imported from Turkey and the Levant, they are generally rather large for ordinary phantom cases; but the little heads produced in our gardens are quite delicate and appropriate. The former should be prepared for maceration by removing the seeds through a gimlet-hole in the stem. The delicate lace-work interspersed among the coarser fibres which give it shape is seldom perfectly preserved; but, if a little opening is found necessary in the side, to remove the interior decayed structure, it will occur to every one, in mounting the specimens, to present the fair side to view.

There is great room for experiment in the use of seed-vessels and mature calyxes, the outer floral envelopes, for this ornamental purpose. Some campanulas, mallows, nettles, and scull-caps have already been rendered tributary to it; and in the English cases considerable prominence is given to the graceful bell-shaped calyx of henbane, and of the scarcely less elegant belladonna, plants, which, though not naturalized in the United States, are cultivated, and sometimes found growing spontaneously in the neighborhood of botanic gardens. From experience in skeletonizing the inflated calyx of the *Nicandra physaloides*, or ground cherry, and of a small cultivated species of purple tomato brought to our markets, they can be recommended as furnishing most delicate additions to the phantom case.

The large heads of the common garden hydrangea yield, by maceration, fine preparations, in which each calyx is distinctly and beautifully skeletonized. These are best mounted in little clusters or as single flowers, attached to slender stems, where their extreme lightness can be displayed to the best advantage.

Bleaching

Next to the perfection of the tissues, their perfect whiteness is most important. This is best secured by soaking them in a solution of chloride of lime, which may be made with water alone in a proportion of from one to four ounces of the chloride to the pint. More delicate leaves and calyxes require a weaker solution than that used for the stramonium and poppy-heads: they are to be bleached in a glass or queensware vessel, and removed in a few hours, or as soon as the requisite brilliant whiteness is reached, then washed off with clear water and laid away to be mounted.

Some sprigs of broom-corn, and the delicate petioles or leaf-stalks which have been separated from the leaves in skeletonizing, should also be bleached, as they will be needed subsequently in mounting the specimens.

Some ferns, lycopods, and mosses—especially when browned by age or etiolated by the deep shade in which they sometimes grow—are easily bleached and adapted to occupy conspicuous places in mounted collections. They contain less fibrous structure than the more highly organized leaves used for skeletonizing, and when the green coloring-matter of their cells is completely obliterated, they are so nearly transparent as not to interfere with the airy lightness at which we aim. They are to be bleached in the same solution used for the skeleton leaves, and carefully floated off on to paper, and pressed between the leaves of a book till needed for mounting.

Labarraque's solution, (sometimes called solution of chloride of soda,) a preparation occasionally used in medicine and as a disinfectant, is preferred by some for bleaching the skeletons, and, according to my experience, is preferable for ferns, lycopods, or any other vegetable structure in which the green coloring-matter has not been previously removed. This green coloring-matter, called by vegetable physiologists chlorophyle, is an organized material, always present in plants grown in the light. Its relations to the life of the plant are not unlike those of the red corpuscles of the blood to that of red-blooded animals. It is hard to destroy this principle by chemical means; and hence, when we desire to bleach a fern or lycopod, as before stated, we shall succeed better after the partial bleaching, produced by time and the slow process of drying. Old ferns, brought as relics from places visited long ago, and which have been laid away between the leaves of a book, generally bleach with facility with either of the chlorinated solutions. Some operators are in the habit of macerating ferns for a few days and boiling them occasionally in the water, by which means the chlorophyle appears to be partially decomposed, and bleaching becomes practicable.

As much disappointment is apt to occur to the inexperienced in bleaching, it is necessary to observe the quality of the bleaching-materials. Chloride of lime should be dry, or nearly so, and should have a strong chlorinous odor, and care should be taken in dissolving it that it becomes thoroughly incorporated with the water by stirring or trituration. The powder should be poured into the water, not the water on to the powder, and after the solution is made it is to be poured off clear, or filtered, before the skeletons or ferns are introduced. The solution should not be prepared until it is wanted. It is even more important that Labarraque's solution should be fresh and properly prepared; it is a more active bleaching-agent than the solution of chloride of lime. When of good quality, it should be diluted with from four to eight parts of water for bleaching ferns or skeleton leaves, but it is frequently so weak as to require scarcely any dilution. It is always important to remove the preparations as soon as they are perfectly bleached to prevent their becoming too soft and tender.

Mounting the Specimens

The information contained in the foregoing pages, besides its general scope, has been specific enough to enable any whose tastes and instincts incline them to attempt skeletonizing to begin with a fair prospect of success. These will naturally desire some hints towards rendering the specimens they have prepared most available for parlor-ornaments. It will hardly be expected, however, that any attempt should be made to enter into detail on this part of the subject.

The English cases, as before hinted, usually contain too many opaque seed-vessels for our fancy; and many of those arranged in this country seem to have been grouped too compactly, with the aim to display a great variety of objects. Our preference is for a loose, airy-looking arrangement of the leaves, rising at the summit almost to contact with the glass under which they are displayed. This glass shade should be abundantly large and of about the same height as its diameter. The cushion that under-lies the skeletons, to our taste, may best be of green velvet, re-calling the image of verdure in contrast with frost.

An opened and inverted stramonium-bur, with its tough and hollow stem, may serve as a substantial pedestal for the airy superstructure of leaves, while the more solid objects of the collection may be interspersed around and upon its base. One of the most appropriate surmountings is a spray of fine fern, its top curving gracefully under the shade, its fronds slightly pendent as they fall on either side the curved stem.

In a model phantom case, arranged by a medical friend, — himself a model naturalist, "humble that he knows no more," — a deli-cate fern, rising to the summit, trembles with electric vibrations

30

on every touch of a silk handkerchief to the glass, while a little tuft of hydrangea-flowers, loosed from its moorings, rises to the top, like a balloon, whenever the unseen electric flash is wakened even by dusting the surface of the shade.

Although the usual form of phantom case is that of a bouquet under a glass shade, the fashion has lately obtained of arranging the specimens in an oval frame behind a convex glass. The background may then be of some rich, dark color which displays the white skeleton leaves in strong contrast. A grouping of seed-vessels at the base of the bouquet allows of the loose ends of the stems being collected and concealed from view; while the leaves and fern-fronds are spread loosely and widely over the enclosed space, allowing an opportunity for the display of taste in their arrangement. The chief objection to this method of displaying the products of the art arises from a deficiency of light, penetrating the recesses of the frame, to give the brilliant effect obtained in the bouquet covered by a transparent shade,—a defect obviated only in part by using a very shallow frame with a deeply-convex glass. When hung against a wall, this should be always placed opposite a window, or in such position as to command the best possible light. It will then form a fitting ornament to the parlor or drawing-room.

Single seed-vessels, as of the poppy, the stramonium, or physalis, or small groupings of these, constitute neat mantel-ornaments. They may be displayed to advantage in a tall wine-glass, or a small vase, preferably under a suitable shade. It was the intention to illustrate the mounting of the specimens by appropriate representations of these methods; but the phantom bouquet is too delicate and full of exquisite detail to be well represented in an engraving; and even photography here fails to do justice to objects which to almost microscopic fineness add a lustre nearly approaching transparency.

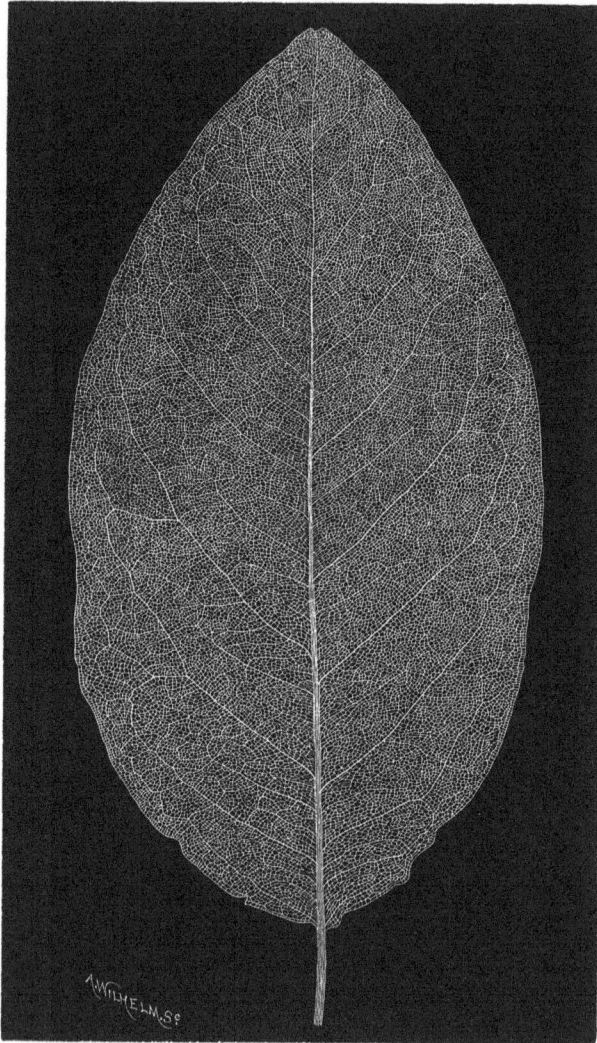

Fig. 4. *Magnolia glauca.*

Aesthetics

In terming the phantom case a perennial source of enjoy-
ment, we have used an expression fully justified by the fact. In
summer, the thoughts it suggests are in strong contrast with
our surroundings: its fleecy whiteness recalls the hoar-frost and
snow-flakes so pleasantly associated with the season of active
exercise and cold and bracing air.

When Sidney Smith conceived the ludicrous notion of so-
lacing himself under the intense heat of the dog-days by laying
aside his too cumbrous flesh, that he might sit in his bones, he
revelled in a fancy which might well have been suggested by
the skeleton case.

On dreary winter days, the bouquet may serve to recall many
a happy hour spent among the trees clad in their summer verdure, — often, perchance, in companionship which has enhanced
the charms of nature, while it beguiled the toil of grateful and
congenial pursuit.

Nor is this fair ornament destitute of that highest function
of nature and art, to lift the soul from grovelling things up to
the regions of poetry and of love. It displays in a most attractive aspect that union of typical forms with special ends which
is everywhere discernible as one of the great initial ideas of the
Creator; it fitly illustrates the secret perfections hidden among
all the grosser material forms which surround us; and, above
all, it is typical of that hidden spiritual outline, obscured by
the grossness of the animal nature, but which, through that love
which is infinite, may survive the inevitable decay, to shine for-
ever in spotless purity and beauty.

Appendix

List of Trees, Shrubs, and Climbers, the Leaves of Which Have Been Skeletonized by Maceration, and of Seed-Vessels Used for the Purposes of the Art.

I. — *Hardy Deciduous Trees and Shrubs*
Maples — including European sycamore.
Poplars — Lombardy, abele, and aspen.
Lindens — American and European species.
Magnolias — including *Magnolia grandiflora.*
Tulip Poplar — *Liriodendron tulipifera.*
Willows — probably all the species.
Beech — *Fagus sylvatica,* gathered early.
Ash — probably several species.
Hickory — the skeleton splits, in drying.
Chestnut — an open texture, difficult.
Horse-Chestnut — *Æsculus hippocastanum.*
Elm — *Ulmus Americana,* "The Treaty Elm."
Kentucky Coffee-Tree — *Gymnocladus Canadensis.*
Pears — The fruit-bearing, and *Pyrus japonica.*
Quince-Tree — *Cydonia vulgaris.*
Apricot — and probably the plum, with care.
Andromeda — an ornamental dwarf shrub.
Deutzia — a remarkably beautiful venation.
Spiræa — several cultivated species.
Sassafras — producing various-shaped leaves.
Althæa — very difficult to prepare.
Pomegranate — the flowering garden-shrub.
Rose Acacia — *Robinia hispida.*

Rose—several species, by long maceration.
Medlar—*Mespilas japonica.*
Wild Cherry—*Cerasus serotina.*
Sugar-Berry—*Celtis occidentalis.*
Witch-Hazel—very desirable: gathered early.
Fraxinella Dictamnus.
Gardenia Florida.
Laurestina.
Franciscea—very handsome.
Erythrina Crystigalla.
Virgilia Lutea.
White Fringe-Tree— *Chionanthus Virginica.*

II.—*Evergreens.*
Holly—difficult to prepare, but desirable.
Mahonia—three species, various forms.
Barberry—*Berberis aristata* and *purpurea.*
Mountain Laurel—*Rhododendron.*
Box—leaves small, with attractive venation.
Butcher's Broom—*Ruscus hypophyllum.*
Olea Fragrans—beautifully serrated.
Camelia Japonica—a universal favorite.
Caoutchouc—*Fagus elastica,* rare.

III.—*Vines and Creepers.*
Ivy—*Hedeora helix,* various leaves.
Bignonia—evergreen.
Wistaria—*Glycena frutescens.*
Dutchman's Pipe—*Aristolochia tomentosa.*
Green Briars—*Smilax rotundifolia* and *herbacea.*
Wild Yam—*Dioscorea villosa.*

IV.—*Seed-Vessels, Modified Leaves, and Calyxes which have been used for the Purposes of the Art, prepared chiefly by Maceration, or found naturally Skeletonized.*
Thorn-Apple—Jamestown-weed, *Datura stramonium.*
Poppy—the Levant heads, and cultivated garden-poppies.

Mallows—several common species.
Nicandra—*Nicandra physaloides*.
Physalis—ground-cherry, *Physalis viscosa*.
Henbane—*Hyoscyamus niger*—English.
Monk'shood—*Atropa Belladonna*—English.
Wild Sage—calyx; various species of *Salvia*.
Safflower—False saffron, *Carthamus tinctoria*.
Canterbury Bells—*Campanula medium*.
Toad Flax—*Linaria vulgaris*.
Skull-Cap—different species of *Scuttelaria*.
Fig-Wort—*Scrophularia nodosa*.
French Tomato—Jerusalem cherry, *Solanum pseudo-capsicum*.
Wild Hydrangea—the mature corymb of *Hydrangea arbor-escens*.
Hydrangea—the calyx of the ornamental species.
Bladder Senna—*Colutea arborescens*.
Bladder-Nut—*Staphylea trifolia*.
Ptelia—Wild Hop—*Ptelia trifoliata*.
False Pennyroyal—*Isanthus cærulia*.

Phantom and Skeleton Leaves
(1883)

Phantom or Skeleton Leaves

Phantom Bouquets, so universally admired by all who can appreciate the chaste and beautiful in art, although but recently introduced to the notice of the American public, are nothing new.

The art of preparing the fibrous skeletons of plants was understood and practiced by the Chinese many centuries ago, and there are still to be found in our fancy stores reasonably perfect specimens of these skeletonized leaves, generally painted and decorated with Oriental designs and mottoes, according to the taste of that remarkable people. Whether they have ever advanced so far as the grouping or arranging of these delicate tissues into anything approaching a bouquet, we cannot say; as no evidences of their faculty for producing such combinations have reached this country; or whether, if they had progressed so far, their stiff and awkward ideas of artistic effect would agree with the cultivated taste of Americans, remains to be imagined.

The works of Chinese art which reach us, whether on lacquered tables, work-boxes, waiters, etc., show how widely their conceptions of beautiful curves and graceful postures differ from our own standards of beauty. But be this as it may, American tourists within the last few years have been struck with the great beauty of these Phantom Bouquets, as exhibited in the fancy bazaars of European cities. These were evidently the work of the few who, in other lands than theirs, had acquired a knowledge of the art. A number of these bouquets thus found their way to this country, where they fortunately came under the notice of cultivated minds, by whom the art of producing them has been so patiently and successfully pursued, that the specimens now produced in this country surpass in richness, brilliancy and faultless nicety of preparation and arrangement, all that have been prepared in foreign lands.

But a few years ago the first Phantom Bouquet ever offered for sale on this side of the Atlantic was made by an American lady, and was exhibited in the spacious window of a large jewelry establishment in one of our chief cities. Although surrounded by flashing silver ware and sparkling gems, yet the little bouquet, composed of only a few phantom leaves and flowers, attracted the highest admiration of all who beheld it, and as may be supposed, it soon found an appreciative purchaser at a very large price. A few others (all that could then be furnished) were disposed of at the same establishment during that season. This public display served to awaken a wide interest in the subject, stimulating inquiry into the wonderful art by which the perishable leaves and blossoms of the forest and the garden are converted into durable illustrations of the complex structure of the floral world.

As is usual with so decided a novelty, many amateurs were ready to experiment the following year. Among numerous lamentable failures, a few only were partially successful in their attempts to reproduce them. We say partially, for in many cases a fine leaf was marred by stains, spots or blemishes occasioned by the ravages of insects; and although otherwise it may have been perfectly skeletonized and the shape preserved entire and beautiful, yet these blemishes served to spoil the effect, and to destroy its value for a bouquet. Many of the less particular artists did not hesitate to mix a few such defective specimens in their arrangements; but most persons of correct taste preferred to group gracefully their half-dozen perfect leaves under a small shade, than to make a towering bouquet of imperfect or discolored ones.

The time which has elapsed since the art was first introduced here has been a season of patient experiment and investigation. There were no published essays to which the learner could refer for directions. All must be studied and acquired by laborious and careful observation, and often whole seasons would be lost while ascertaining the peculiar properties of a single leaf, the process being too slow to allow of a second gathering before Autumn had stripped the trees.

The first summer of the writer's experiments was lost in vain attempts, and bushels of carefully gathered leaves were wasted for want of a few items of knowledge, which to a careless operator, would seem of small importance. Five years of practice have taught her many things indispensable to a successful prosecution of the art, such as are neither understood nor appreciated by those who have just commenced the work. It is the object of these pages to furnish plain and practical directions for producing perfect Bouquets of Skeleton Flowers, together with a list of such plants as will repay the artist's labor.

A late writer on this subject enthusiastically declares that the art is yet in its infancy, and expresses his belief that diligent experiment will lead to results even more wonderful than any that have yet been achieved. In the confident belief that such will be the case, we shall feel glad to have given our readers an impulse in the right direction, and can assure them that by closely following the rules here given, success will certainly reward their efforts. Those whom repeated failures may have so far discouraged as to induce them to abandon the pursuit, will be stimulated to renew their interesting labors. Others, whose entire ignorance of the process may have withheld them from even beginning, will he induced to make a trial. The probability is, that among the aspirants thus stimulated to enter the field, some superior genius will be found, at whose animating touch this beautiful art will receive a brilliancy of development surpassing all that could have been imagined by those who pioneered it into public notice.

Preparing Leaves and Flowers

When Spring has once more dressed both tree and shrub in their gorgeous livery of green, the artist begins to look around her for the most suitable subjects for experiment. The influence of the new study on her mind becomes immediately apparent to herself. The trees, which have heretofore appeared to her as presenting an unbroken uniformity of foliage, now display their leaves to her sharpened observation with a wealth of capabilities before unknown to her, and she is surprised to learn how infinite a variety exists in the vegetable world; variety, not only in size and outline, but in those other characteristics which are so important to her purpose, strength of fiber and freedom from blemishes occasioned by the destructive ravages of insects. As observation is directed to the subject, so the mind becomes expanded under the influence of the new study. The surprising difference between the leaves now first becomes apparent. They are seen to be serrated or entire, ovate, acuminate, cordate or irregular. The magnificent luster of the Ivy and Magnolia now, for the first time, attracts attention and secures for them a new admiration. As the season advances, she will be struck with the numerous changes to which the leaves are subject before the chill winds of Autumn strip them from the trees, thence depositing them in rustling piles upon the ground. As incidental to the study, the habits of a multitude of insect depredators will be noticed, affording new subjects for surprise and fresh accessions of knowledge. Everywhere the wonders of the Divine Hand will be displayed under conditions to which she had been a stranger; and the mysteries of Nature thus unfolded will infinitely surpass all we may mention in these pages.

Without some directions to guide her, the enthusiastic learner, in haste to begin the work, gathers indiscriminately from forest and garden, selecting leaves remarkable only for their ample size or pleasing shape, and places the whole diversified collection in the prepared receptacle to undergo the process of maceration. In her ignorance of certain first principles, she does not imagine that she has overlooked some of the most indispensable ingredients of success, which, standing as they do at the very threshold of the undertaking, must not only influence, but when disregarded, must render absolutely futile, in subsequent steps in a process which under any circumstances is exceedingly tedious. We may suppose that in her natural impatience to commence her labors she has gathered up an ample store of leaves, immediately on their attaining their full growth. It is true that in this early preparation she has anticipated the attacks of destructive insects, but the leaf will then be too immature to withstand the macerating process. The fibers will be found too succulent and not sufficiently ligneous to sustain the pressure and handling always necessary to produce a perfectly skeletonized leaf. After probably two months of patient watchfulness, she is consequently compelled to throw away her choice collection, the whole having become a mass of pulp, in which there is neither stem nor fiber to identify a single leaf.

By this time the season has advanced and the foliage on the trees has undergone important changes. Many of the leaves, having lost their early succulency, have assumed a strong ligneous character. In place of excessive pulpiness, an undue proportion of fiber pervades the whole structure of the leaf. It has, in fact, become too old for maceration. In other cases the leaf has either been stung by an insect, and the channels through which the sap so mysteriously circulates having become obstructed by the poisonous infusion injected into them, its shape becomes distorted, or its surface is disfigured by blisters. Others have been attacked by a different tribe of enemies, who by half devouring the leaf, as effectually destroys it for the artist. The latter catastrophe invariably overtakes the foliage of the Elm, the Magnolia and the Maple. These facts we have verified in our own experience; and having been compelled thus to learn them, the resulting knowledge was acquired only from repeated and trying disappointments. They make evident the importance of knowing time exact point in the season at which each leaf is in proper condition for the artist's hand.

Another error consists in placing in the macerating vessel many different sorts of leaves, without a knowledge of their chemical properties. For instance, those of the Oak, Chestnut, Walnut, Birch and Hickory contain so large a quantity of tannin as to render it almost impossible to macerate them in the usual way. If placed among other and more perishable leaves, the infusion of tannin thus created will act as a preservative and entirely prevent their decomposition. The writer learned these facts, to her cost, during the first season's experiments. A few beautiful Oak leaves were placed among a large number of other varieties which were in course of preparation, and not until after months of patient waiting, watchfulness and handling did she discover the true cause of her disappointment, when it was too late in the season to repair the loss. The reader will at once perceive how important are these rules and cautions, thus placed at the head of our directions.

Throughout the Middle States by the fifteenth of June most of the desirable leaves will be found fully grown, and many of them are then old enough to gather. Elms, Swamp Magnolias, Maples, Deutzias, Pears, Silver Poplars and English Sycamores may be selected, but none but the firmest and most perfect leaves should be taken. These kinds may be placed together in open vessels and covered with soft water, and then set in a warm or sunny place in the open air. A broad weight may be placed on the top, so as to insure continued immersion. A newspaper, doubled and laid over the top of the leaves, will answer the same purpose as a weight and is perhaps better, as it keeps its place while the weight sometimes falls to the bottom of the vessel. The best vessel for the purpose is a common earthen jar with a wide mouth, the size to be proportioned to the quantity of leaves to be macerated.

At the end of six weeks the paper may be removed and a few of the leaves carefully taken out for examination, and placed in a basin of clean warm water. To do this, the human hand is the best instrument; but as many persons may object to thus dipping into what has now become an unpleasant mass of vegetable decomposition, a broad wooden spoon may be substituted. Then, taking a leaf between the thumb and finger, immerse the hand in the warm water and press and rub the leaf either gently or thinly, according to the strength of its texture. This rubbing process will remove the loose green matter from the surface and expose to view the fibrous network of the leaf. With those which are strongest, especially the Swamp Magnolias, a brush will be needed to effectually clean them—a soft toothbrush will answer best—but in using a brush, the leaf should be laid in the palm of the hand, on a plate, or on any other surface equally flat and smooth.

This constitutes the first washing, and a few of the leaves will now be found perfectly clear. But to some of them thus washed and but partially cleared further care must be extended. It will therefore be necessary to have at hand a second vessel of water similar to the first, in which all such imperfectly skeletonized leaves may be placed, where they must remain until finished, which, with all but the Swamp Magnolias, will probably be two or three weeks longer.

We may suppose that the artist has made a beginning with the leaves already mentioned in this chapter. On taking them out of the macerating vessel and washing them as directed, she will find the Deutzias and Silver Poplars perfectly clean, and they should then be placed in a basin of clean water until all the contents of the macerating jar has been examined. A few of the Norway Maples will also be found perfectly prepared; but the majority of all contained in the jar will still be only partially so.

In the latter condition will be found the Sycamores, the Silver Maples, the Elms and the Pears. These must, consequently, be deposited in the second vessel, as before mentioned, to undergo still further maceration. The Magnolias will require another two or three months' soaking before the outer cuticle will become soft enough to remove; but if more convenient, they may be placed in the same vessel with those last named. After covering these half-cleaned leaves with water, all in different stages of progress, they should be left in the same warm, sunny place to be finished. We may here remark, for the comfort of the learner who has persevered thus far in an operation which will be discovered to be decidedly unpleasant to her olfactory organs, that the most offensive portion of the labor is over, at least with this particular set of leaves, as after having received their first washing, they part with most of the putrefactive odors which have so long pervaded the air in the vicinity of the macerating jar.

The clear and perfect leaves which were deposited in the clean water, awaiting a leisure hour to give them further attention, may now be deprived of their moisture by carefully pressing them between the folds of a soft blotter until they are perfectly dry. On no account let them be laid on a table, or other hard surface, while in a wet state, as in drying they will adhere to it so closely as to tear in the effort to remove them. The Norway Maple, being extremely delicate, will adhere, while wet, even to the hand, and great care must be exercised in removing its leaves to avoid tearing. It will be noticed that many of the leaves will lose their stems in passing through the process; but the mode by which this deficiency is to be supplied will be explained in its proper place hereafter. When dried, the leaves may be placed in boxes, ready for bleaching when the assortment has been completed.

We append another method, which may not be so efficient, but which is more expeditious and not at all offensive:

First dissolve four ounces of common washing soda in a quart of boiling water, add two ounces of slacked quick-lime and boil for about fifteen minutes. Allow the solution to cool; afterwards pour off all the clear liquor into a clean saucepan. When this liquor is at its boiling heat, place the leaves carefully in the pan and boil the whole together for an hour, adding from time to time enough water to make up for the loss by evaporation. The epidermis and parenchyma of some leaves will more readily separate than those of others.

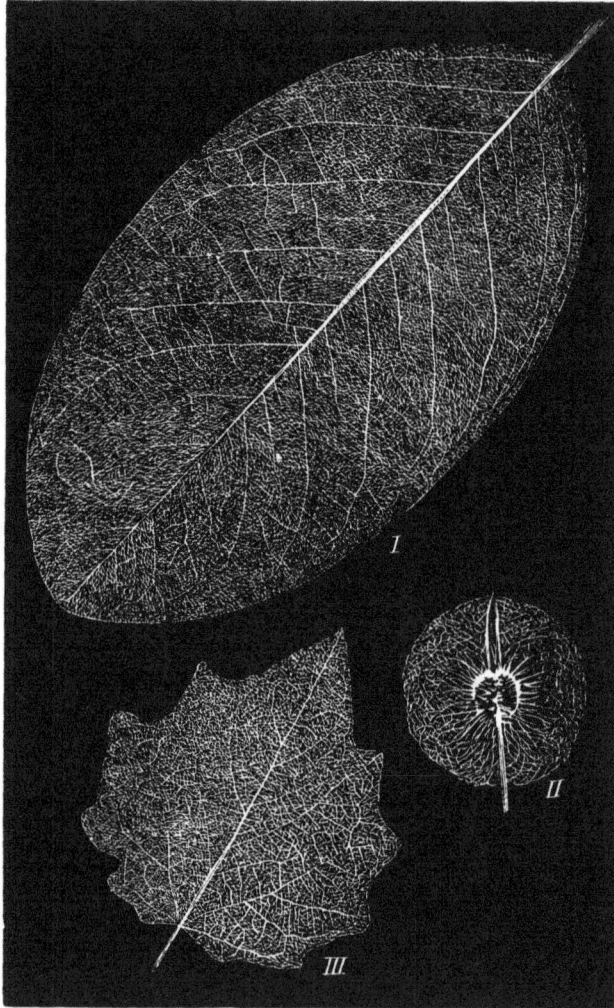

Bleaching Leaves and Seed Vessels

The next process, and one of great importance, is that of bleaching the leaves, flowers and seed-vessels. It is an operation which requires the greatest care, as upon the perfect whiteness of all the component parts of a bouquet its beauty will depend. No matter how perfectly the leaves and seed-vessels may have been skeletonized, if they are permitted to retain any shade of their original yellow they are deficient in beauty, at least to the eye of the connoisseur.

The first step in this part of the process is to procure proper bleaching materials. Many persons are entirely successful in the use of chloride of lime, while others prefer Labarraque's solution of chloride of soda, or Powers & Wightman's. The former should be prepared for use in the following manner: Take a half pound of strong chloride of lime and place it in an earthen or other pitcher. Add three pints of soft, cold water and stir carefully with an iron spoon, pressing so as to mash the lumps well against the sides of the vessel. Keep it covered and allow it to stand in a cool place until the lime has precipitated upon the bottom of the pitcher, which will be done in about an hour, except a small portion that may remain floating on the surface. This should be removed with a spoon or skimmer, after which the clear liquid should be poured off into a bottle, then corked up tightly and kept in a cool place.

When ready to commence leaf bleaching, take a glass jar, such as is used for pickles or preserves, having a mouth wide enough to admit the largest leaf. First, select those intended to be whitened, but be careful not to place leaves and seed-vessels in the same jar; then with soft, clear water cover the leaves in the

jar and add the bleaching solution, which is extremely power-
ful and should be diluted with from three to six times the quantity of water (soft), according to the texture of the leaves to be
bleached. The jar should be covered tightly and set in a warm
place. When coarse seed-vessels and stems are to be bleached,
this proportion of the chloride of lime may be doubled, but the
delicate leaves, and especially the Ferns, will be destroyed if
the solution be made too strong.

Labarraque's preparation of chloride of soda acts gently and
more slowly, and being free from the caustic properties of the
lime, is less likely to attack and corrode the delicate framework of
the leaves. The quantity of this solution to be added to water must
be double that of the first named preparation. It will whiten the
flowers, Ferns and more tender of the seed-vessels, but it is not
strong enough to act on those winch are coarser and more ligneous. There is great difficulty, however, in procuring this preparation of the required freshness and strength, as its bleaching properties depend entirely on the amount of chlorine contained in it;
and this being a very volatile gas, it is readily lost by keeping a
length of time, even when carefully corked and sealed.

The best preparation for this purpose is Powers & Wightman's. One bottle of this will whiten a large number of leaves,
without injuring the fiber or making them brittle, as is the case
with the chloride of lime. The proper proportion for mixing will
be about half a teacupful to a pint of water. This will generally
whiten two sets of leaves; that is, as soon as those first put in
are perfectly white, they may be taken out and a second lot
placed in the same mixture. Sometimes, however, it will be necessary to add a small quantity more, say a tablespoonful, in
order to complete them. For amateurs, and even for accomplished artists, a superior solution, thus ready prepared will
be found safer and more likely to insure perfect success than
any preparation they will be able to compound for themselves.
The saving of trouble in using it will be quite a consideration.

In putting the delicate leaves into the jar, care should be
taken to arrange them beforehand with the stems all pointing
the same way, that is, downwards in the jar. The reason for this

exists in the fact that the bleaching commences first at the bottom of the vessel; and as the thick stems and mid-ribs require more time to whiten than the lace-like portion of the leaves, it insures their being satisfactorily finished in a short time. A jar of leaves will usually require from six to twelve hours for bleaching; but as the jar is of glass, an outside inspection will enable the operator to judge of the degree of whiteness without raising the lid until it may be time to remove them.

When they are discovered to be entirely white, they must be taken carefully out with the hand and laid in a basin of clean, warm water. If suffered to remain too long in the jar they will become too tender for removal. They may then be thoroughly washed from the chlorine, by changing them several times in fresh water, after which they will be ready for their final drying. This is accomplished as before, by laying them between blotting pads; while the more delicate ones, which are apt to curl in drying, should now be laid between the leaves of a book until entirely dry. The washing is a very important part of the operation, as if not thoroughly done, the bouquet will soon become yellow and otherwise discolored, and thus in the end lose its attractiveness and beauty as a parlor ornament.

As before stated, it will be advisable to keep the seed-vessels separate from the leaves and to put them in different bleaching jars. If placed promiscuously in the same jar, the seed-vessels will become so entangled in the fine network of the leaves, that in the attempt to remove them the latter will be seriously injured. Seed-vessels and flowers require the same treatment in bleaching and washing, only remembering that the coarser seed-vessels may need a stronger infusion of the bleaching preparation. A little experience will soon inform the operator as to the exact quantity required for all kinds of leaves and seed-vessels.

The bleaching of the Ferns will need some special directions. Many who have succeeded admirably with leaves, have invariably failed in their attempts at preparing these graceful sprays. As they constitute the most brilliant embellishment which can be introduced into a bouquet, such failures are especially mortifying. But by closely following these simple directions, there will be no difficulty in producing entire sprays of white Fern ready to be arranged with other materials for the bouquet.

Having gathered Ferns of different varieties during their season of maturity—which is when the seeds are to be found on the back of the leaves—they should be preserved by pressing them between the leaves of a book, there to remain until required for bleaching. When ready for that process, let the operator select such as she desires, and place them carefully in a jar, causing them to curl around the sides rather than with stems downward, in order to avoid breaking the dry and brittle leaves. The smaller separate leaflets may occupy the space in the center of the jar. Then fill up the jar with warm water, leaving room for the bleaching solution, in the proportion of half a teacupful of the solution to a pint of water. Cover the jar tightly and set in a very warm place. After twenty-four hours, gently pour off the liquid and replace with fresh, mixed as before. They should remain in the second water about forty-eight hours, when this, in like manner, will require to be changed. In about three or four days the Ferns will begin to whiten at the edges, and this whiteness will gradually extend itself over the entire surface of the spray, changing it from a dark, brownish green to the spotless purity of a snowflake. Each one must be carefully taken out as soon as it is seen to be entirely white, without waiting for the whole contents of the jar to be finished.

In the bleaching of a large spray, it sometimes happens that its extremity, perhaps half of the entire length, will become perfectly white, while dark spots remain on the upper or stem end. In such cases it will be safest to take out the branch, and laying it in a basin of water, cut off the white portion, and return the unfinished remainder to the jar. Afterwards, when both are ready for the bouquet, the two portions can be neatly united with gum arabic. The process of changing the water will have to be repeated four or five times during the operation of bleaching the same lot of Ferns, and the time required to whiten them completely will extend over a period of from one to two weeks. The time depends on the varieties of Ferns which may be used, as there is a wide difference in their susceptibilities, some being wholly unfitted for this purpose.

When the sprays are found to be entirely white, they must be taken from the jar with the fingers, always holding them by the stem, and laid in a broad basin of clean, warm water, where they should be allowed to remain for several hours. They may be thoroughly rinsed by changing the water several times, but they will not bear handling in the same manner as will the skeleton leaves. When ready to be dried, take one spray by the stem and lay it in a broad dish or basin of water, allowing it to float on the surface; then pass under it a sheet of unsized white paper, and in this way lift it out of the water. The spray will cling to the paper, and assume its natural shape. Should any of the small side leaves become crooked or overlapped, they may be readily straightened by using the point of a pin to spread them out in proper shape upon the paper. To get rid of the superfluous moisture contained in the latter, lay the sheet first on a soft blotter for a few minutes. The blotter will absorb most of the excess of water. After that it must be laid between two other sheets of the same unsized white paper, and pressed in a book.

When all the sprays have been thus removed and committed to the keeping of the book, a heavy weight should be placed on it, in order to insure their drying smoothly. If desirable, the drying may be accelerated by changing them, after a day or two, into another hook, or into new portions of the first. When entirely dry, if some of the thinner varieties are found to adhere to the paper, they may be loosened by pressing the thumb nail on the under side of the paper. It is better, however, even after they are thoroughly pressed and dried, to keep them shut up in a book until wanted for the bouquet, as they have a tendency to curl when exposed to the air.

The writer has given directions for the bleaching of Ferns only by the new preparation of Powers & Wightman, as it has been proved to be the most reliable compound for that purpose. She has fully tested chloride of lime, and finds it altogether too severe for these delicate tissues, while Labarraque's solution is much slower in its operation—one bottle of the new preparation being equal in strength to two of the article last named.

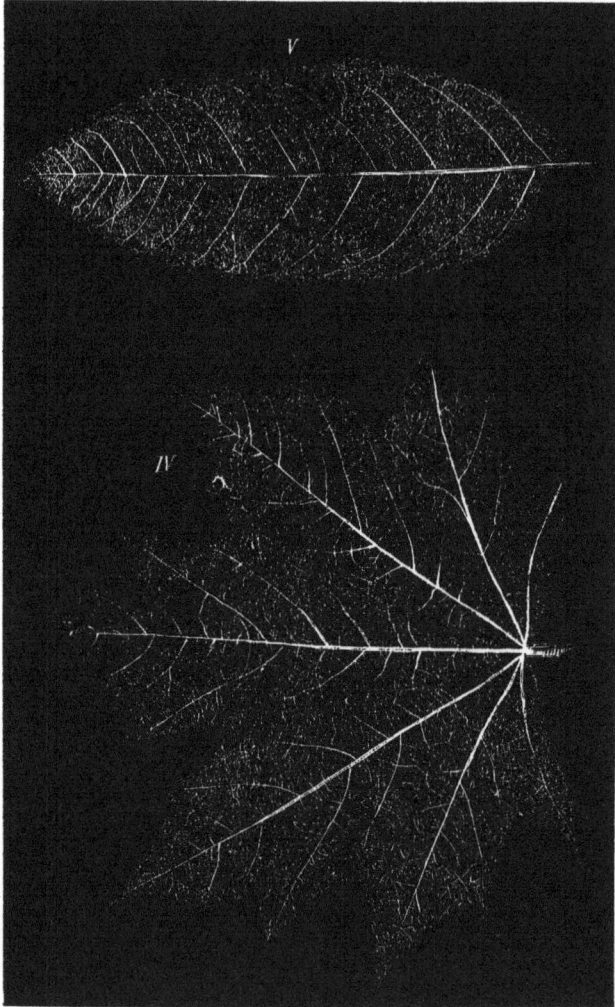

List of Plants for Skeletonizing

The process of maceration has already been shown in the preceding pages. We have endeavored to give such clear and practical directions as will apply to all varieties of plants, but there are certain peculiarities which seem to be inherent in each particular leaf, seed-vessel and flower, so as to call for specific directions, in order that success may be insured with all. Instead, therefore, of dismissing the subject with a mere list of leaves adapted to the purposes of the art, and leaving each learner to discover these varying peculiarities for herself, at great cost of time and labor, we shall give a few general rules for the treatment of each one named. The learner will need all the light that can be thrown on the subject, and the minute particulars which follow will contribute largely to her successful prosecution of the art. The illustrations which accompany the description of such leaves as are most important, will enable the reader to determine the names of doubtful varieties.

Magnolia
This splendid genus of trees deserves to be placed at the head of our list of those plants whose leaves are well adapted to the purposes of our art. Its varied species are to be found on the eastern shores of both the great continents of North America and Asia. The United States produces no less than eight varieties, while China and Japan have four or five. Neither Europe, Africa nor South America can offer a single species of indigenous Magnolia.

The different varieties of Chinese Magnolia have, with one or two exceptions, been acclimated with us, and are to be found in most of our ornamental shrubberies, their lovely white and purple blossoms and spicy fragrance, together with the neat and regular appearance of the trees themselves, making them general favorites. Most of the Chinese varieties will answer for our purpose, but we give preference to the following: First,

White Chinese Magnolia
(Magnolia conspicua)

This variety blossoms during April in the Middle States, and by the Chinese is called the Lily Tree, from its lily-shaped flowers of a creamy white color. The leaves arrive at perfection in June, and may be gathered for maceration between the 15th of that month and the middle of September. After that time the ravages of insects begin to show themselves.

Magnolia Purpurea and Magnolia Soulangianna are purple varieties of Chinese origin, and may be gathered and treated as the above-named. From four to six weeks will generally be long enough for their perfect maceration, when they can be readily cleaned by the aid of warm water and rubbing between the thumb and finger.

American Swamp Magnolia
(Magnolia glauca)

(Fig. No. 1.) This is the fragrant wild Magnolia, which blooms in June, and is found in great profusion in the swamps and marshes of New Jersey. When transplanted to the garden the leaves are produced in great perfection, while their size is increased by cultivation. They are in perfection at the time of blossoming, and on no account should be gathered later, as after that time they become too tough and abound with invisible stings of insects, which injuries, not becoming apparent until after the cleansing process has been completed, the otherwise beautiful leaf will be found covered with small black spots which can neither be whitened nor removed. These leaves require three or four months to macerate, and may then be brushed with a tooth-brush to remove the little cellular particles which fill up the interstices and which give to them a thick and cloudy appearance.

Silver Poplar
(*Abele*)
(Fig. No. 3.) This leaf is one of the most desirable, as well as most easily cleaned since it requires but four or five weeks to macerate, and has a strong fiber. The leaves of this tree present much variety of shape, and the sizes of those which are matured vary from half an inch to four inches in length. They may be gathered as early as the 1st of June, and generally remain free from spots until September. Avoid the foliage of the suckers, which are frequently found growing vigorously around the parent tree, as the fibers of such leaves are too weak and tender for our purpose. They will lose their stems by maceration, but these may be replaced, as directed in a previous chapter.

Aspen Poplar
(*Populus tremula*)
The leaf of this tree is larger than that of the preceding and is also more delicate. It may be gathered in June or July, and will require about a month to macerate. Great care will be necessary in handling them.

Tulip Poplar
(*Liriodendron tulipifera*)
Lombardy Poplar
(*Populus pyramidalis*)
Both these may be gathered early in summer, and should be treated like the Aspen Poplar.

Norway Maple
(*Acer platanoides*)
(Fig. No. 4.) The most beautiful of the Maple family in shape and general adaptability to the present purpose. A single branch taken from one of these trees will present great variety in size and shape, the small leaves at the extremities cleaning quite as perfectly as the largest. They should be gathered by the 20th of June, certainly not later than the middle of July. They will be finished in about six weeks, losing their stems, as is invariably

the case with all Maples. The Silver Maple may be treated by
the same rule.

Lindens and Weeping Willows

(Fig. No. 16. Willow.) These two desirable leaves may be
gathered in July, and will macerate in from six weeks to two
months. They need very careful handling, or brushing with a
camel's hair brush on a plate.

European Sycamore
(*Acer pseudo-platanus*)

(Fig. No. 6.) A beautiful leaf, in shape somewhat resembling
the Norway Maple but possessing a firmer and thicker texture.
It must be secured early in June, as by the close of that month it
becomes unfitted for our use, and but few of those collected
after the 20th of June will come out entirely free from clouds or
blemishes. About two months will complete their maceration.

Ash

(Fig. No. 5. English Ash.) There are several species of this
family which are admirably adapted for our object. Of these,
the Flowering Ash (*Ornus Europaeus*) and the English Ash are the
most beautiful. They will become clear and perfectly skeletonized in about six weeks after gathering, which may be done in
July and August.

Everlasting Pea, or Chickling Vetch

(Fig. No. 11.) This pretty garden perennial, with an abundance of deep pink blossoms, is too well known to need description. The leaves may be gathered at any time during summer and require but a few weeks for maceration. They lose their
stems. The graceful tendrils of this vine may also be placed in
water with the leaves and after remaining some weeks the outer
cuticle can be easily removed without untwisting the curl, and
these, when bleached, will be found ornamental to the bouquet,
especially where the design adopted consists of a vine.

Elm

The leaves of this beautiful tree must be gathered very early. Indeed, so soon do the caterpillars begin their ravages, that in some sections of the country, before the leaf is strong enough for the purpose of the skeletonizer, it is too much eaten to be worth collecting. June or July will answer, if any perfect leaves are then to be found. They will macerate in about four weeks, and, being very delicate, will need the greatest care. If the leaf be laid on a plate, or something similar, a camel's hair pencil will remove the softened particles, leaving the fiber clean, to be floated off into the basin of water, and then laid carefully on a towel to dry.

The Evergreen Elm (*Ulmus sempervipens*) (Fig. No. 10) is a small, glossy leaf with scalloped edges and may be used at any season of the year, requiring about three months for its perfect clearing. A native of France and is rare in America.

Deutzia Scabra; or Rough-Leaved Deutzia

(Fig. No. 8.) One of the most beautiful small leaves we can use. Gather in June or July. They will be perfectly skeletonized in three or four weeks, without losing their stems. These graceful little leaves, with serrated edges, form beautiful wreaths and sprays, either for black velvet crosses or to be twined around the base of a bouquet.

Deutzia Gracilis, another variety of this desirable garden plant, requires somewhat longer for its perfect preparation.

Beech, Hickory and Chestnut

These leaves contain a slight portion of tannin and had better be kept separate from other kinds. A few drops of muriatic acid added to the water in which they are placed for maceration will hasten the process. They may be gathered in July and will require several months to become completely skeletonized.

Dwarf Pear, Sassafras and Althea

(Fig. No. 9.) Gather iii July. They require about two months to macerate.

Rose

(Fig. No. 7.) The common annual blooming dark velvet Rose
furnishes the best description of leaves for our purpose. They
should be gathered in July before the insects have stung them,
and will require about two months' soaking. They are very delicate and must be brushed on a plate.

White Fringe Tree
(*Chionanthus Virginica*)

Gather in July. Will be ready for clearing in about two
months.

Dutchman's Pipe
(*Aristolochia tomentosa.*)

This is a rather coarse vine, of rank growth, well suited for
covering unsightly buildings or decaying trees. It bears a curious white blossom, shaped somewhat like a pipe, whence it
takes its homely name. The leaves are heart-shape, of thick and
woolly texture, but the skeletons they produce are so exceedingly beautiful as to make them indispensable to a complete
collection. They should be taken from the vine not earlier than
the middle of July, and perfect specimens may be obtained as
late as the middle of September—probably about the first of
August will be the best time. Select the firmest and oldest leaves.
Some of them will be clear in four weeks after immersion.

Ivy

(Fig. No. 17.) These much admired leaves may be gathered
at any time during the year, always selecting those a year old
in preference to the younger growth of the present season. The
Ivy leaf, like some others, has a tough outer cuticle on each side,
between which the fibrous skeleton is concealed, the interme-
diate space being filled with the green cellular matter common to
all leaves. During the process of maceration this green substance
becomes dissolved, though the outer skin remains whole and
entire. When taken from the macerating vessel and laid in the
clean water for cleansing, this skin will present the appearance

of a bladder filled with green water. By puncturing, or gently tearing the skin on one or both sides of the leaf, the water will escape and the perfect skeleton will float out, ready for rinsing and drying. Four or five weeks will be sufficient to allow for their preparation, although some varieties require a few weeks longer.

Holly

(Fig. No. 12.) This leaf is quite difficult to clear properly, owing to the tough outer cuticle adhering so tenaciously to the thorns on the edges, as to tax the ingenuity and patience of the operator in removing the one without breaking off the other. For this reason most amateurs give up after the first attempt and content themselves with more beautiful and less trouble-some subjects. About three months is the time necessary for skeletonizing them; and being evergreens, they may be gath-ered at any time.

Wisteria, Bignonia, Greenbrier and Wild Yam—all vines that are tolerably well known—may be skeletonized by the usual process in from six weeks to three months, and should be gath-ered about the middle of July.

Of greenhouse plants, the leaves of Camellia Japonica, Cape Jasmine, Laurestina and Caoutchouc may be done after months of soaking. A shorter process, however, which some parties pre-fer for all descriptions of leaves to the slower method which we have adopted, is found to answer well for these particular species. Their tough epidermis requires something more than the ordinary sluggish operation of water and summer heat to soften and remove them. The process consists in boiling them for several hours in strong soapsuds, using the ordinary chemi-cal soap of the shops.

This will generally succeed with these last named plants, but for those which are tender and delicate, as before described, it is too severe. Besides this, the chemical properties of the soap affect the leaf in so peculiar a way as to increase the difficulty of bleaching; and notwithstanding all possible care be taken to wash after the boiling process is over, enough of the refractory element remains to defeat all attempts to make the leaf perfectly and permanently white. Therefore, while we mention the process as an item of information due to the learner who desires to understand the whole routine, and to test for herself the various modifications of practice now in use, yet we prefer and still adhere to our own formula, as at first described. We consider it the best, and by far the most reliable, although it is unquestionably slow and tedious in all its various processes.

In concluding our list of these, the most desirable leaves that have so far come under our own observation, we would by no means limit the researches and experiments of other artists. Different localities will unquestionably furnish different specimens, and thus their collections may be greatly enlarged by the adoption of new and more beautiful leaves. As a general rule to govern in the selection of appropriate subjects for experiment, let those of strong and woody fiber be chosen, rather than thick, fleshy leaves, whose veins or ribs may be soft and juicy. Avoid, also, those which have veins traversing the leaf in a longitudinal direction, instead of forming a network tissue radiating from the mid-rib to the outer edges of the leaf. The former are known as endogenous, the latter as exogenous varieties of leaf structure. As an example of the endogenous, we may cite the leaves of different kinds of Lilies. If put into the macerating vessel, a few days, or a week, will be sufficient to reduce them to a mass of pulp, resembling a bunch of thread or strings, with apparently no connecting framework to hold the fibers together in form. The practiced eye can in most cases discover the character of the leaf under observation, by merely holding it up against the light, when the veinwork will be plainly perceptible, and its value decided by the closeness or coarseness of its vascular structure.

We add the following as having been successfully skeletonized:
Horse Chestnut (*Æsculus hippocastanum*).
Kentucky Coffee Tree (*Ginnocladus Canadensis*).

Flowering Pear (*Pyrus Japonica*).
Andromeda.
Rose Acacia (*Robinia hispida*).
Witch Hazel (*Hamamelis Virginica*), said to be very beautiful; should be gathered early.
Wild Cherry (*Cerasus serotina*).
Sugar Berry (*Celtis occidentalis*).
Fraxinella Dictamnus.
Franciscea, —very beautiful.
Erythrina Crystigalla.
Virgilia Lutea.
Matronia.
Barberry (*Berberis aristata*, and *purpurea*).
Mountain Laurel (*Rhododendron*).
Box.
Butcher's Broom (*Ruscus hypophyllum*).

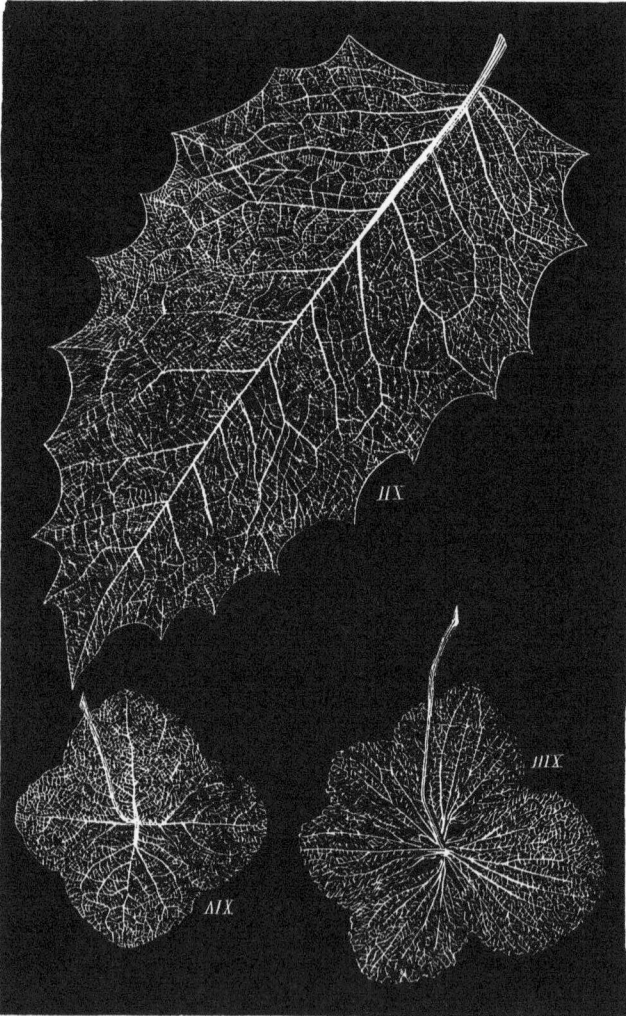

Seed Vessels

Different varieties of the Ground Cherry family (*Physalis*) are entitled to particular notice. The peculiar characteristic of this family of plants is the berry, enclosed in a bladder-like receptacle. These berries are about the size of the cherry, with color yellow, red or purple, and having a pleasant, sweet taste. The green covering becomes of a yellowish color when the fruit is ripe, and they fall to the ground together, when the curious case will soon become perfectly skeletonized by contact with the damp ground. But as they are very liable to be eaten by insects while on the ground, it is much better to gather them as soon as they fall and place them in the macerating vessel, allowing the berry inside to remain until softened, in order to avoid tearing the delicate little bladder. Two or three weeks will be long enough to allow for their preparation. They may be washed by passing rapidly to and fro in hot water, when the softened berry may be pressed out, then dried with a soft blotter. Some species lose their stems and may be prepared for the bouquet by using the gummed thread, being careful to bend gracefully, so as to give the effect of drooping.

Wild Hop
(*Ptelia trifoliata*)
(Fig. No. 2.) This is a membranous capsule surrounded by a leafy border, which after about two weeks' soaking, becomes very lace-like and beautiful. Before bleaching, the seed may be removed by making an incision on one side of the capsule, being careful when afterwards arranging it, to place that side downwards.

Nicandra Physaloides

One of the most desirable and showy for this purpose. The blue Nicandra should be cultivated by all makers of the Phantom Bouquet. The calyx of the plant, enclosing first time flower and afterwards the seed capsule, is of a curious balloon shape, of bright green until the seed is ripe, when it becomes brownish. Each one has a tough stem, which is retained through maceration, and is attached to the stalk of the plant, the latter being covered by the calyxes, at a distance of an inch apart, quite to the end of the branch. This calyx seems to be formed of five distinct divisions, like leaves, which, when pressed open and bent in proper shape, has after bleaching, every appearance of a flower. To increase the variety in the bouquet, they can be used both in their natural form to represent buds, or in the way described. They require about three weeks to macerate, when they may be cleaned in hot water, aided perhaps by the toothbrush. A whole branch may be done without separating from the main stem.

Thorn Apple: Jamestown Weed
(*Datura Stramonium*)

A well-known rank wayside weed, very poisonous to the taste, but not to the touch. The seed-vessels should be gathered when ripe, and soaked about six weeks, when by the aid of a stiff brush, the beautiful skeleton will appear. When bleached, they resemble carved ivory, and are much admired in the bouquet. The only drawback to their value is their tendency to become brown again after bleaching. For this reason we have entirely discarded them.

Wild Cucumber, or Balsam Apple
(*Echinosystis*)

This is one of the most curious specimens in our list of beautiful seed-vessels. It is said to grow in abundance in the neighborhood of Boston, bearing a profusion of seed. The seed-vessels vary in size from an inch to nearly two inches in length, and about half that in thickness. They become perfect skeletons on the vine, where they should be allowed to remain until the frost has opened them and dropped the seed. If not entirely clear when gathered, they may be completed by a few weeks' soaking. They form beautiful vase-like receptacles for the base of the bouquet, and as they retain their whiteness, are excellent substitutes for the Stramonium burrs.

Lobelia

The little wild species is very beautiful, with its delicate globes set along the stem. About three weeks will do for them, when they will become clear by passing to and fro in a basin of hot water.

Skull Cap
(*Scutellaria*)

These delicate clusters of seed-vessels may be skeletonized in two or three weeks, and cleared in the same way as the Lobelia.

Shell-Flower
(*Phyton Concha*)

A curious shell-shaped calyx, with four seeds which remain in the extreme point of the horn. The plant is rare and rather difficult to cultivate. It seems to belong to the Sage family, and has an aromatic odor when pressed. The calyx is very delicate, and will macerate in ten days or less. When seen in a group of Phantom Leaves, they somewhat resemble the Convolvulus blossom.

Poppy

The cultivated garden varieties will macerate in a week or two. The fiber does not remain very perfect, at least in a general way, as it is apt to tear by removal of the inner skins. But the star-shaped summit of the capsule looks well upon the velvet cushion. The black lines which radiate from the center may be removed by aid of a pin, when a beautiful lace-work appearance will be imparted to it.

Mallows
Several varieties. The common garden Mallows, with calyx enclosing seeds, are the prettiest. They grow in clusters, and if suffered to remain until a frost, will become skeletonized on the plant.

Hydrangea Hortensia
(Fig. No. 13.) The well-known garden species—the bunches should be left on the plant until late in September, in order to become firm. Separate into small bunches, leaving not more than four or five in a cluster. They will require about ten or twelve weeks for maceration, and may then be cleansed by passing to and fro in hot water, changing the water frequently as it becomes filled with loose particles. If some of the leaves are separated, they can easily be replaced with gum arabic after bleaching.

Hydrangea Quercifolia: Oak-Leaved Hydrangea
(Fig. No. 14.) This is a tougher and coarser species, composed of four flat petals. It requires longer time to macerate than the Hydrangea Hortensia, but should be gathered as soon as the bunches begin to turn brown on the tree. A brush will be necessary to clear properly.

Campanula
The seed-vessels of the several species of these, including the Canterbury Bell, are much admired in the bouquet, although not so delicate as the Lobelia, which they resemble. Some varieties will become sufficiently prepared on the plant and only require bleaching, but others require two or three weeks' maceration.

To the above list the following may be added:
Black Henbane (Hyoscyamus niger).
English Monkshood (Aconitum Napellus).
Toad Flax (Colutea arborescens).
Wild Salvia.

Figwort (*Scrophularia nodosa*).
Jerusalem Cherry (*Solanum pseudo-capsicum*).
Bladder Nut (*Staphylea trifolia*).
Safflower: False Saffron (*Carthamus tinctoria*).
False Pennyroyal (*Isanthus cerula*).
Lily of the Valley: The dried Flowers.

In concluding these instructions in the art of preparing and completing the Phantom Bouquet, we have endeavored to be plain and practical in every particular, seeking not only to direct the learner in her experiments, but also to guard her against the mistakes and disappointments which must invariably attend the labors of the unassisted amateur.

When the first bouquets appeared for sale in this country, the admiration they excited awakened a general curiosity as to the process by which they were produced. Inquiries were addressed to the editors of some of our scientific journals, but they could answer only according to their own very limited knowledge of the art; and hence this occasional information was exceedingly vague and unreliable, and, indeed, it often misled the learner, resulting in discouragement to some and in entire disgust to others.

The writer has here given her own practical rules and ideas, adopted from actual experience, and no careful learner need hesitate to follow in her footsteps. But, however invaluable instruction may be to the beginner, personal experiment will be found indispensable. We cannot write up the amateur to the position of an artist. Yet a desire to reach the status of the latter will stimulate to exertion and perseverance, and these, with ordinary taste and skill, will surely be rewarded with success. None, therefore, whose love for the truly beautiful in art is deep and strong, and whose aspirations for eminence are decided and sincere, will permit a few early discouragements to turn them aside from the undertaking.

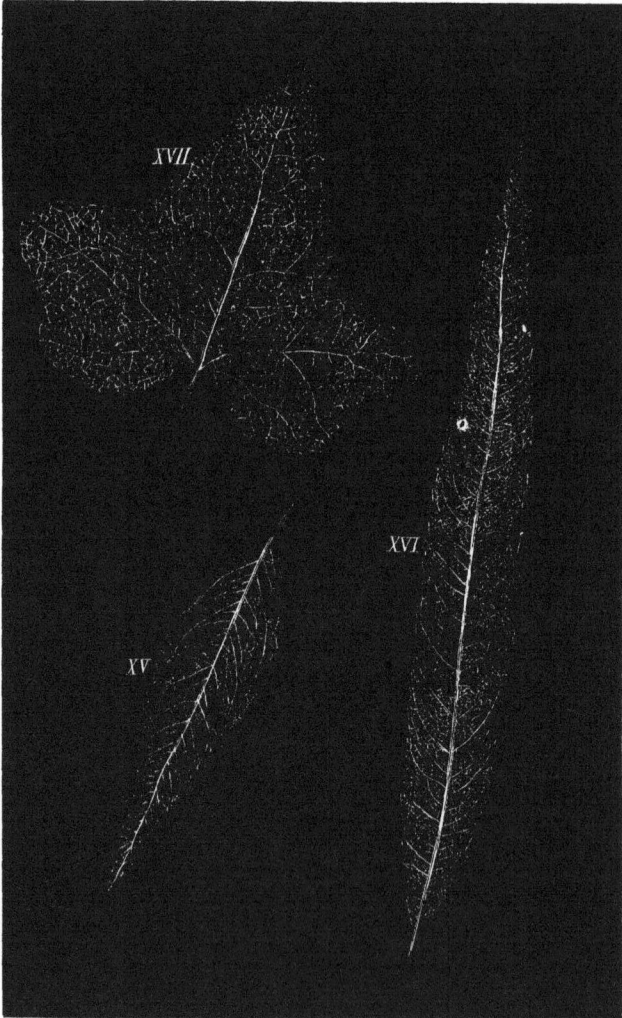

Leaf Printing

Many times it will be desired to make impressions of the skeletonized leaves, either for preservation as curiosities in the scrap book or photograph album, for transmission by mail as specimens of the art, or for the engraver to reproduce on wood. The making of these impressions directly from the leaves, though an exceedingly simple process when once understood, requires much care and skill to learn. Whoever may undertake to produce them, should call in, if possible, the aid of some friend who has a practical knowledge of printing, as the processes by which books and newspapers are printed are all applicable to leaf printing.

The operator should procure a spoonful of printer's ink and with a case-knife spread a small quantity over half the surface of a marble slab about a foot square. When spreading the ink on the slab, let it be confined to one end of it, not letting it cover more than half the stone. Care must be taken not to allow thick streaks or ridges of ink, but to spread a thin film or covering as uniform as possible. As printing ink is a thick and paste-like compound which stiffens in cold weather, if the operation is to be performed when the temperature is low, the stone should be slightly warmed before the ink is laid on. The warmth will render the ink sufficiently fluid to operate in a satisfactory manner. If no marble slab can be conveniently obtained, then a smooth board, about an inch thick, may be substituted. The board will not require to be warmed.

When the stone has been supplied with ink as above directed, a roller is passed several times over it, until the whole surface of the roller becomes coated. It will take up the ink in unequal

71

quantities—that is, more in one place than in another—with just as much irregularity as it has been laid upon the stone with the knife. This irregularity must now be remedied, and the ink distributed over the entire surface of the roller with absolute uniformity. This is quickly accomplished by frequently passing the roller to and fro over that half of the stone on which no ink has been spread. But in so doing, care must be taken to occasionally lift it from the stone and to give it a half revolution before again putting it down, so that its surface shall come in contact with new portions of the surface of the stone. By following these directions the ink will become distributed evenly over the surface of the roller, whence it will be transferred with corresponding uniformity to the delicate framework of the leaf, and will produce a perfect impression of its most complex vein-work. If the ink is not thus nicely distributed on the roller, the interstices in the leafy structure will become filled with it and the impression will present an unsightly blotch.

For taking impressions, thin letter paper will be found the best, if it be nicely glazed and free from ridges or water-marks. It should first be cut into pieces about the desired sizes, and then slightly sprinkled with clean water, say two or three pieces first. On these as many dry ones should be laid, and they sprinkled in turn, then more dry ones, then another sprinkling, and so on until the whole quantity has been sprinkled. Let the pile lie for half an hour, or until the paper has absorbed all the water.

Then take the pieces, one at a time, and turn them over, placing the first on a board, and the others on top of the first, but shifting them about as they are turned; that is, if a very wet end or corner is observed in one piece, turn the piece around so that the excessively wet places shall come in contact with dryer surfaces in the new pile. Be particular to smooth all wrinkles with the back of the thumb nail. If the paper has been made too wet, the accident can be remedied by interposing dry pieces between two wet ones. When the whole has been turned, put a slight weight on the pile to press all down smooth, as much depends on having the paper in perfect order.

Being now ready to commence the printing, a leaf is placed on a smooth board, with its under side uppermost, as there the leafy veins or ribs are more prominent than on the upper side. The roller having been charged with ink, it is rolled to and fro over the leaf until the latter is seen to have received a sufficient supply. Three or four times going over will generally be enough. Then lay the leaf with the inked side down, on the top sheet of the damp paper pile, and over it place a double sheet of dry paper, press on with the left hand so tightly that the leaf shall not move, and with the thumb nail of time right hand rub pretty hard over the whole leaf. This pressure of the thumb nail will transfer the ink on the leaf to the surface of the damp paper, and if the inking has been carefully done, a clear and distinct impression will be obtained. All the leaf impressions contained in this volume were taken for the engraver by the process described above.

Modern Methods for
Skeletonizing Leaves

Modern Variations

The historical methods for creating skeleton leaves will certainly work, but modern variations are worth experimentation. Generally, simmering leaves in near-boiling water speeds up the process, but as noted in the historical sections, this has to be done carefully, or may damage the leaves. Some sources suggest drying out the leaves first, by inserting them within the pages of an old telephone book and leaving them in a warm dry spot for a few weeks.

Once the leaves are ready for simmering, you can add 1 teaspoon of baking soda and 1 teaspoon of baking powder per quart of water. (You can also experiment by adding 1 to 2 teaspoons of washing soda to the water, instead. Just remember to use gloves, as it can be caustic.) Add the leaves to the water, heat to boiling, then reduce to a simmer, stirring gently until the leaves are soft and pulpy. This can take up to 40 minutes. They can then be carefully strained out, rinsed in cold water, and the pulp removed with a small brush (or rubbing by hand, with some leaves).

After the leaves dry, they can be placed in a bleach solution for up to two hours. You can experiment with the amount of bleach, but try 1 to 2 tablespoons per quart of water to start. When bleached, rinse the leaves well and press them dry in paper towels for at least two hours.

After the leaves are bleached and dried, they can be arranged or framed. You can also color them by placing them in a mixture of rubbing alcohol and food coloring.

Coachwhip Publications

CoachwhipBooks.com

www.ingramcontent.com/pod-product-compliance
Lightning Source LLC
Chambersburg PA
CBHW022127280326
41933CB00007B/585